I0532334

SOCIAL MEDIA FOR ACTORS

100 Essential Tips to Grow, Thrive and Survive Online

HEIDI DEAN

Copyright © 2024 Heidi Dean
All rights reserved.
No part of this book may be reproduced or used in any manner without
the prior written permission of the copyright owner, except for the use
of brief quotations in a book review.
To request permissions, contact the publisher
at Marketing4Actors.com

Paperback: ISBN: 979-8-9899975-1-0
E-book: ISBN: 979-8-9899975-0-3
Library of Congress Control Number: 2024903011

First Paperback Edition: February, 2024

Editor: Philip Michael
Copy Editing by: Liz Dexter
Cover art by: Rocko Spigolon

Heidi Dean
New York, NY
Marketing4Actors.com

The advice and strategies found within may not be suitable for every
situation. This work is sold with the understanding that neither the
author nor the publisher is held responsible for the results accrued from
the advice in this book.

All brand names and product names used in this book are trademarks,
registered trademarks, or trade names of their respective holders. I am
not associated with any product or vendor in this book.

DEDICATION

To my mom and dad for always supporting my dreams.

To my brother, Kip, who supported my dreams even after I dressed him up as a raindrop so I could sing "Don't Rain on My Parade" (your Oscar for Best Supporting Brother is in the mail).

To my dearest Tallulah: you can do anything you put your mind to.

To my amazing husband, Phil, for always making me coffee. Without you, hundreds of videos, livestreams, and this book would never have happened.

And to every actor who ever had a dream. You are never too old to pursue what sets your heart on fire. Thank you for allowing me to cheer you on, every step of the way.

TABLE OF CONTENTS

Author's Note

Hi! It's Heidi. It's a little weird to see you on paper here. I'm used to seeing you on social.

This book took me five years to write. Well, not exactly five years, but it took me that long to wrap my head around how I could write a book about a topic that changes every day. #TrueStory.

But, the more I pondered it, the more I realized there are core strategies that have endured across platforms, through pandemics and algorithm changes. These strategies are the reasons for my early success starting on Twitter (when it was still called Twitter). Then I replicated that success on Instagram, cultivated a thriving Facebook Group, ventured into YouTube, and even rocked it on TikTok. It didn't matter what year it was or which platform I was on, these strategies worked.

Now, I'm sharing them with *you*. It's time for you to unlock your potential and find success on your social media journey.

I've also created a companion guide to this book to help you #DoTheWork. You can use it to take notes but, more importantly, the guide makes it possible for me to keep you posted on my most up-to-date strategies and app recommendations. In it you'll also find an additional 100 tips, exercises, and strategies! #YoureWelcome. You can access it for free at www.socialmediaforactors.com/guide.

Before we dive in, we need to take a moment to talk about why social media matters for actors. You might be reading this book because your agent or coach told you it was important and advised you to improve your social media presence. But it's easier to commit to something when you understand WHY it matters. There are many reasons but the main ones are:

1. **It's becoming part of your job.** Social media used to be a tool for savvy actors to build relationships, but now it's often included in contracts. Your next gig could have social media posting guidelines or requirements, live social media Q&As, or even an account 'Takeover.' Are you prepared?

2. **It's part of your first impression online.** People are constantly searching for you online. In this industry of referrals, whether they go directly to your social media profiles or do a Google search for your name, your social media accounts appear at the top of the search results. What they find needs to look professional. We'll explore this further in this book, I promise.

3.   **Social media helps you build relationships.** The most important reason an actor needs social media is for building relationships. Before social media, relationships were limited to who you could meet in person. But now you can connect and build relationships with the people you want to know online. Don't worry; I've dedicated 15 of the 100 tips in this book to this topic.

4.   **It can help you get cast.** Yes, having a real, engaged online audience can open doors for you (more on that in Chapter 12). But it goes beyond just followers. In an industry where actors have limited power, social media offers you a virtual stage from which you can share your talent with the world. *Social media gives you control over your career.*

Now, I can tell you these things, but if you haven't shifted your perspective to recognize the enormous career opportunity social media represents, listing a hundred more reasons won't make a difference. So, before we dive in, let's have a little #RealTalk… Social media is a powerful tool only if you see it as one.

After working with thousands of actors, I've noticed there are basically two types of actors on social media. Imagine they're both looking at a path of stones.

Actor A sees the stones (social media) as **stumbling blocks**, something they *have* to do: "Why do I have to be on social media? It's a waste of time. I just want to act."

Actor B sees them as **stepping stones**, as an opportunity to share their talents, grow their audience, meet more people in the business, and strengthen the relationships they already have.

Nothing in this book will work if you see the stones as stumbling blocks. You'll never get good at something you hate or you don't want to do. So, if Actor A looked eerily familiar, you've gotta switch your mindset and start seeing social media as a stepping stone, not a stumbling block. Your social media success (and your career) depends on it.

Let's do this!

Heidi Dean

CHAPTER 1

OPTIMIZING YOUR FIRST IMPRESSION

Today, casting happens fast and it starts online. In this digital age, you'll probably make your first impression online before you ever meet someone in an audition room.

This means your social media profiles must directly reflect who you are and the kind of person you'll be to work with on set, on stage, or in the booth. If you want to use social media to create more opportunities for your career, your first impression has to make a *positive* impact.

In this first chapter, you'll find twelve quick fixes for improving your first impression online.

IMPORTANT: Do not skip this section. Neglecting to clean up your first impression is the biggest reason you won't see results with your networking and audience-building efforts.

Be Consistent With Your Profile Photo

For most actors, this will be your headshot or a professional photo from your website (or the red carpet) that captures your essence and vibe. Make sure the photo you choose is cropped close to your face and used consistently across all of your social media profiles. One of the most common mistakes actors make on social media is using different profile photos on different platforms. It can take people between 7 and 12 times to recognize you online. So, make it easy for them to know it's you, no matter where they find you online.

Think of your profile photo as your logo. Just like McDonald's doesn't use different logos at different locations, the same applies to you. Your profile photo should remain consistent across all your social media platforms.

Valuable Real Estate: Your Header Photo

On platforms (like Facebook and YouTube), we have a large header at the top of our profile. This valuable piece of real estate is an excellent place to display who you are and what you do as an actor.

Not sure what to include in your header photo? Try one of these ideas:

- Keep it simple: A photo of you and your name.
- Highlight a project: A still from your favorite project.
- #BookedIt Billboard: Announce a booking in your header.

Don't forget to check the companion guide to this book at www.socialmediaforactors.com for my favorite apps that'll help you create professional header photos for free.

Claim Your @Usernames

Your professional stage name is your brand. Make sure you claim @yourname across social media, even if you don't think you'll use that platform right now. If you have a common name, it's quite possible someone already has your username. So here are some alternatives:

- Use part of your name
- Use your initials
- Shorten your name
- Add actor, actress, acts, VO, Voiceover, Narrator, etc.
- Add Real
- Add The
- Add Iam, Im, or Its
- Create a memorable phrase that tells us what you do: @HeidiOnScreen @VoicesByHeidi

Have another username hack not listed above? Connect with me on Instagram (I'm @Marketing4actors) and let me know!

Did You Change Your @Username?

After you change your username, update the following:

- Your email signature
- Website social icons and follow buttons
- Links in your social media bios
- Your business cards (if they have your @username)
- Your IMDb/Online casting profiles

You don't want broken links across the internet.

Four Tips For Writing Your Bios

Your social media bios play a pivotal role in your social media success. First, when someone is deciding whether or not to follow you, they'll check your bio. Second, when someone (including a casting director) Googles you, your username and bio appear in the search results. If your bio is poorly written (or non-existent), your efforts to network and grow your following could be for nothing.

Here are three things to include and one thing to leave out of your social media bios:

1. **Include Keywords** that attract like-hearted people and those who want to hire you. There are some wonderful social media bios by celebrities out there— use them to inspire you, but don't forget to add keywords. Reese Witherspoon's bio doesn't

need to explicitly state she's an actor, but the world might not know who you are... yet. Be sure to mention "actor," "voiceover actor," "audiobook narrator," so we 'key in' to how you fit into the business.

2. **Include a Value or Interest** (outside of acting). Saying what you do in the business is awesome, but why not go further by telling us what you care about. Are you passionate about the environment? Animals? History? Or horror films? This will help people connect with you and *want* to follow you.

3. **Add a Connection or Accomplishment.** Your alma mater, acting school, representation, theater company are all possibilities. Choose your own adventure. If you've been nominated for or have won an industry-related award (like Sundance, Jeff, Tony, Audie, etc.), include them in your bio.

If you're just starting out and don't have a lot of professional experience, your bio should still sound accomplished. Please **don't** say you're an "aspiring actor." If you're taking acting class, submitting for projects, auditioning, and working on your career, you're an actor. Period. No 'Aspiring' needed. Don't let anyone tell you otherwise.

Pin A Post!

The first few posts on your social media accounts must clearly represent who you are and your talents! Why? These are the first posts everyone in the industry will see when they land on your profile.

Most social networks let you 'Pin a post' (or multiple posts) to the top of your profile. This ensures that everyone who visits your account sees those posts first.

Not sure what to pin? Here are some ideas:

- Your latest booking (if it's aired/announced)
- A post of you performing
- An introduction post (that tells us who you are and what you post about)
- Your best-performing post or favorite post
- A post with your current Headshot(s)
- A post highlighting your favorite acting roles

Common Questions When Starting Out:

What Social Networks Should I Be On?

Choose your platforms based on your career goals, not just where your friends hang out.

Ask yourself:

- Will this platform help me reach my career goals?
- Do the industry pros I know (and want to know) spend time here?
- Do fans of the show you're in spend time here?
- Is this a platform I actually enjoy?

P.S. These are also great questions to ask before you decide to jump on the new platform everyone's talking about.

Once you're done answering them, you may realize that a Facebook Page isn't helping you reach those goals because it's not helping you with your networking and it's draining your energy. Instead, maybe you just focus on your Instagram for a little while. Listen, it's better to totally rock one platform than to spread yourself thin and get burned out on four. Choose the one(s) that are right for your career and your goals and give them 100%.

Pin A 'Party Post'!

If you've decided not to play on a platform, add your profile and header photos and optimize your bio. Then, pin a post that says you're not active on that platform and link them to where you do 'party.' There's an example of a 'party post' in the companion guide to this book.

"Should I Have A Public Or Private Account?"

In most cases, if you're using your social media accounts for your acting career, remember that an actor is a public figure. You need a public account.

Here are four reasons why a private account is NOT the way to go if you want to reach your goals:

- Growing your following becomes nearly impossible with a private account. To hit that follow button people need to see what you're posting about. With a private profile they can't.
- Want to post about your projects to help spread the word? Unfortunately, your reach is severely limited with a private account.
- Considering using hashtags with a private account? Think again. Hashtags won't be effective because you won't appear in hashtag feeds for new connections where you could pick up potential new followers. In fact, you won't show up on search and discovery pages anywhere on social media.
- Even with a private account, your followers can still save and screenshot your posts. Nothing is completely private on social media.

Instead of having a private account, find a way to tell your story online, on *your* own terms. We'll discuss this in Chapter 2.

#HomeworkTime

If you know me, you know I'm all about #DoingTheWork. So, before you go, pinky promise me you'll put into practice what you've learned from Chapter 1 and make your first impression rock!

CHAPTER 2

"WHAT DO I POST ON SOCIAL MEDIA?"

Every day, I get questions like: What should I post on social media? How often should I post? When should I post about my bookings?

Over the next four chapters, we'll delve into the biggest mistakes actors make with their posts. These mistakes hinder your ability to grow an audience, connect with industry professionals, and post consistently. Let's begin with the not-so-simple question: "What do I post on social media?"

You Are *More* Than The Projects You Book

If I were to check out your social media, would I hear your authentic voice and understand your story? Or would I just see your latest booking?

Actors who really rock their social media have a nice balance of professional and personal posts. A big mistake actors make is thinking that because their social media is for their acting career, their posts should be "all acting - all the time."

And YES, posting about acting is important (as long as you're not spoiling anything with these #bookedit posts—more on that in Chapter 3).

But there are two big problems with this 'acting only' posting strategy. First, not even the biggest actors work all the time. So if you only post about your #actorslife, you're eventually gonna run out of things to post (or post nothing but #ThrowbackThursdays).

Again and again you're gonna be asking, "What the heck do I post?!"

Second, an 'all acting, all the time' account only tells us half your story. Your story is so much more than the projects you book.

Your followers don't just want to know you, the actor; they also want to know the person they're following. In fact, they want to know how you're just like them. And as you grow your audience, most of your followers won't be actors. So yes, give them a backstage pass to your #actorlife, but also give them a backstage pass into *who you are*.

The Spaghetti Account

Most actors struggling with 'What to post' fall into one of two categories: the 'all acting - all the time' account or the polar opposite, the Spaghetti Account.

While the 'all acting - all the time' account is so narrowly focused that it doesn't let your followers in enough, the spaghetti account is too general and tells us way too much. It's like every time you post you're throwing spaghetti at the wall, hoping something sticks. Sometimes your posts are about acting, but then you share some things you love, your beliefs, tidbits about your friends, your dinner, your running adventures, your cats, your garden, and a sunset. Unless you're a celebrity, spaghetti accounts usually result in very low engagement, numerous unfollows, and minimal benefit for your acting career.

And the more random topics you post about, the more challenging it becomes to build an audience for your content. The social media algorithms (we'll discuss those in Chapter 5) struggle to pin down what your account is about, making it harder to identify and reach people who will appreciate your content. So, if you've ever wondered, "How do I grow my following?" the answer lies in getting specific about which parts of your life you'll post about.

Aim to strike a balance. We're not just posting about acting and we're not delving into a gazillion random topics.

The Magic Posting Equation

If you can share your #Actorslife plus a few values/interests that are important to you, you'll stand out in a very crowded online world and form connections that lead to loyal followers who stick around (and tell their friends about you).

If you've only been posting about acting, I understand it might feel a little scary to put yourself out there. But you're not putting your entire life online. You're just leaving little pieces of your story for us to connect with.

You'll find 25 post ideas at the end of this book to help you tell your story online.

Think Beyond Your Bookings

Actors don't work all the time. That's just part of your #actorlife. But being between gigs doesn't mean you've got nothing to post.

In this chapter you learned that your story is more than the projects you book. It's also about sharing your *journey* as an actor and striking a personal/professional posting balance by incorporating some of the things you love and value.

So, if it's been a while since your last booking, you might consider posting about:

- What you do every day to become a better actor
- The classes you're taking
- What you learned today during your self-tape session

- Your special skills
- What books you're reading
- The performances that inspire you

All of these things are entirely unrelated to your bookings. When you make that switch, you'll discover you have a wealth of posting ideas, whether you're booking or not.

#MindsetMoment

REPEAT AFTER ME:

My story matters. I am empowered to share it with people who want to hear it.

Don't be afraid to share it with the world!

CHAPTER 3

POSTING ABOUT YOUR PROJECTS

To Post Or Not To Post? THAT Is The Question

Before hitting the post button, always ask yourself three important questions:

1. Does this respect my contract or NDA (if you've signed one)?
2. Does this reveal any spoilers about the project?
3. Does this post reflect my *best* self?

When you've booked a job, review your contract or NDA (if applicable) for any restrictions on what and when you can post. Restrictions can apply as early as the audition stage if you've signed an NDA.

If it's permissible to post, move on to question #2: Does this reveal any spoilers about the project?

Finally, consider whether your post portrays your authentic, BEST self, not an 'Insta-perfect' self. Strive to showcase the professional actor that others will want to collaborate with on set or in rehearsal.

Everything Is Content

If there are no specific guidelines prohibiting you from taking photos, capture some quick photos or videos while you're on location. Often, actors become so engrossed in their work that they leave without any media to help generate buzz about the job. Don't let that be you.

Here are some photo ideas to consider on your next job, even if you don't plan to post them until it's allowed:

- Snap a photo of yourself in hair and makeup.
- Take a picture in your trailer, the recording booth, or the dressing room.
- Capture a photo near the project or character name.
- Document photos or videos in costume.
- Share your pre-set ritual or warm-up routine.
- Get shots from the red carpet, cast party, or wrap party.
- Take behind-the-scenes photos or videos with fellow cast members.

With every item on this list, always be aware of your surroundings to ensure you're not inadvertently revealing any spoilers.

Not sure if you should post it? My dad taught me, "it's always better to be safe than sorry." If it's not explicitly mentioned in your contract, ask the production team, or simply take behind-the-scenes photos and wait to post them until after the show airs.

Turn Off Your Location

VERY IMPORTANT: When posting from set, ensure that your location is turned off on social media. This could inadvertently reveal details about where you're shooting and what you're working on, even if the photo and caption don't.

Social Savvy Google Tip

Set up Google Alerts for "Your Name" and the projects you're involved in, and you'll receive press, articles, and announcements about those projects via email as soon as they happen. This provides you with valuable content that you can share on social media to support the project.

Queue Up Your Posts

If you've ever taken photos on set or backstage during rehearsals, but you weren't able to share them right away, you're not alone. Waiting until your show airs, premiers, or opens is just the 'promotional waiting game' that's part of your #actorslife.

Instead of waiting to create these posts and risk these photos end up getting buried and lost in your camera roll, I recommend using a social media scheduler to prepare your #Bookedit photos in advance. A scheduler can queue up posts from your bookings, complete with edits, captions, and hashtags. Then, when your show airs, opens, or premieres, all you have to do is hit the post button.

There are lots of social media schedulers to choose from. In the companion guide for this book at www.socialmediaforactors.com/guide, I recommend my favorite, affordable schedulers.

CHAPTER 4

EFFECTIVE SOCIAL MEDIA PROMOTION

Disclaimer: As discussed in Chapter 3, always refrain from posting about any acting project unless it has aired or you have received explicit permission to share.

Posting Without Boasting

Social media is a powerful tool for promoting your acting projects and content *if* you use it wisely.

When you book a job, share the news with pride and power—it's a win! Share exciting updates with us and create some buzz. Remember, people follow you on social media for a reason and anyone who isn't supportive of your success is someone you should be happy to see unfollow you. Bye-bye!

But, there's a fine line between celebrating your bookings with pride and continuously boasting about them. In this chapter, I'll introduce you to some savvy self-promotion tips that will help you generate buzz without annoying any of your friends or followers.

Social Media Is NOT A Megaphone

Social media is social—it's in the name. It's meant for engaging *with* your audience, not talking *at* them. Too many actors treat it like a megaphone, broadcasting their news at us instead of using it like a telephone to talk with us.

Think of your social media more like a telephone than a megaphone.

It's all about having conversations—a dialogue. When you turn your posts into conversations, especially the promotional ones, you can share about your projects and bookings more frequently. Why? Because people respond better when you shift the focus from "All about me, me, me" to "WE!"

Make It A Dialogue

The simplest way to transform a post into a conversation is to turn it into a question. Instead of saying, "Watch my web series" in the 10th post this week, why not ask, "What's your favorite episode this season?" Now, it doesn't seem promotional at all; it's a conversation starter. Asking a question prompts a dialogue with your followers. Making your post a conversation has a great side effect, too—it will increase engagement on your posts.

Post With Gratitude

When posting about a project you worked on, instead of talking about yourself, invite someone else into the conversation by making them the subject of your post. The subject of your posts can be your followers, fans, a fellow cast member, the director, or the whole project.

For example, you can say, "It was such an honor to work with _____," and tell us what you learned or how they've impacted your career.

'Posting with Gratitude' works every time if you do it authentically, because you're posting about the people you've worked with and the post is about them, not you.

Posting with gratitude has a wonderful side effect too – it's a great way to reconnect with people you've worked with.

Replace These Words In Your Posts

When I come across your post in my feed, whether it's a conscious thought or not, I'm always thinking, "What's in it for me?" Shifting your language from "Me" to "We" will naturally pique my interest every time.

Review your posts and identify words like "I," "me," and "my."

Whenever possible, replace them with "you" and "your." This simple change instantly makes your promotional posts less self-centered.

Get Creative! Have Some Fun!

Think about your favorite TV commercials. They're entirely promotional, designed to sell a product, yet they often make us laugh, evoke emotions, and inspire us. They have a way of making us forget that it's a promotion. Similarly, using an audio trend on short-form social media content (like Instagram Reels) to add humor while promoting your project can be a clever approach to "posting without boasting." If you need help with this type of content, the companion guide has some valuable resources.

#HomeworkTime

Take a look at a few of your recent #BookedIt posts and:

- Find opportunities to start a conversation by turning statements into questions.
- Find places to change your language from me or my to you or yours.

CHAPTER 5

ANSWERING YOUR BIGGEST POSTING QUESTIONS

I've been asked pretty much every question about posting on social media. In this chapter, you'll find my answers!

"Do I need to post every day?"

Before you dive headfirst into posting daily because your friend does or some random YouTube video recommended it, consider two crucial factors:

Your Career

You're an actor, and that should always be your top priority. Between self-tapes, acting classes, your bookings, and your personal life, ask yourself, "How much can I realistically post without it taking over my #actorslife? Without it affecting my mental well-being? Without burning out?" Is it three times a week, once a week, or every day? The answer is different for everyone.

Also, the majority of social media advice you stumble upon, whether on YouTube or the Explore page, is not tailored to actors. It's often aimed at businesses that have the resources to hire a social media manager to post every day.

Your Goals

The other critical factor to consider is "What are my social media goals?"

A. Building meaningful relationships
B. Sharing your talents with the world
C. Supporting the projects you're involved in
D. Growing your audience

If your goals align with A, B, or C, there's no need to post every day. Period. Even if your goal is D (audience growth), you still need to find a consistent posting schedule that is right for you, and that may not necessarily mean daily posting. Consistency is more important than frequency.

"I'm busy! Do I need to engage with my comments?"

Social networks aren't just about posting; they're about connecting and engaging. Right after I post, I schedule ten minutes of social media management time. I check my messages and notifications and then I acknowledge the first people who comment. You have to care about your followers before they ever care about you. When your followers see you responding, they feel a personal connection to you and it makes them feel seen and special. The more you engage with your followers, the more engaged they'll be. Please stop posting and ghosting!

"When is the best time to post on social media?"

The next time you come across an article proclaiming "The Best Time to Post on Social Media," consider taking a detour in the opposite direction. Is there a universal best time to post on social media that applies to everyone? Nope. Could there be an optimal posting time specifically tailored to you and your audience? It's entirely possible.

Here's a little secret: You won't find the answer in blogs or YouTube videos. Instead, it's hidden in your Analytics* (sometimes referred to as 'Insights') within your social media account. Your analytics hold a treasure trove of insights about your followers, such as:

- The days and times they are most active online
- Their age range
- Their gender
- Their location

This invaluable information can guide you in deciding when to post and assist you in crafting more engaging content.

*Note: Analytics are typically available in Business or Creator Social Media Accounts, rather than personal profiles.

"How do I write captions that actually sound like me?"

Remember working on a monologue when you didn't know who you were talking to? It probably made you feel (and look) unfocused, right? Well, when you don't know who you're talking to on social media, that's how your posts come across—totally unfocused.

To solve this problem, choose a real friend and write to them every time you post. Don't just pretend to write to them; actually write as if you are talking to them in person. This helps you write authentically. Experiment with writing to different friends or family members until you find your unique voice.

Timesaving Tip: Use the 'Dictation' button on your phone to speak your captions. This eliminates the need for manual typing and ensures that your captions sound like you.

"How do I get more engagement on my posts?"

One of the easiest ways to increase engagement on any social network is to incorporate a call-to-action (CTA) into your posts and videos.

A CTA is an invitation for your followers to take a specific action after viewing your content. Here are some examples of effective CTAs:

- Tap ♥ if this resonates with you.
- Share this with someone who needs to see it.
- Save this post for later.
- Like this post if you feel the same way.
- Make some noise in the comments if you agree!
- Pose a question that encourages people to comment. There are many ways to do this but it could be as simple as "Whaddaya think?" or "Has this ever happened to you?"

If we don't ask our followers to take action, they usually won't. So, make it a habit to encourage engagement with every post.

"What hashtags help me grow my following?"

Let's shift our perspective a bit. Instead of viewing hashtags as growth tools, think of them as *conversations* between like-minded people on social media.

Keep in mind the more popular your hashtags are, the more people are in the conversation. That will make it more difficult to find *your* people on the platform. The solution: Getting more specific with your hashtags. Start by researching these two types of hashtags:

- **Community Hashtags**: Think about who you want to be a part of your online community. Are you a #womanincomedy, #womaninfilm, a #theateractor, an #audiobooknarrator? Maybe you're a #deafactor or an #Asianactor. Are you deeply entrenched in the #horrorfam or do you proudly wear the badge of a #shakespeareactor or #shakespearenerd? Your combination of 'Community Tags' is completely unique to you. So, consider who your community is and identify the conversations you want to engage in when people search for these tags.

- **Location Hashtags**: These are valuable no matter where you're located, but they can be especially helpful if you're outside a major market. Consider hashtags like #ChicagoActor, #MiamiFilm, #Atlantaactors, #PortlandActor, #LAactress, #AussieActor, or #VancouverActors. These hashtags help you connect with others in the industry within your region.

Take a few moments to pinpoint your specific Community and Location Hashtags. By doing so, you'll strategically position yourself in online conversations that align with your goals.

"Is it okay to share other people's content on my social media?"

Sharing other people's content that aligns with your interests is fine as long as you give them credit. Many of the social networks give us features to help us do this (Ex: share, retweet, remix, or duet). But don't let reposting be a replacement for an actual posting strategy. When I visit your feed and only see reposted content from others, it doesn't give me insight into your story or perspective. It's essential to share *your* thoughts and opinions from time to time so we get to know you. Don't be a reposting robot.

"The algorithm doesn't like me!"

If I had a dollar for every time I heard this from an actor I'd be a gazillionaire!

First, let's define what social media algorithms are. Social media algorithms are sets of rules and formulas that decide what content you see and when you see it. They examine your activity, who you follow, what you watch, and how you engage, in order to curate content they think you'll enjoy.

If you're struggling to appear in your followers' feeds and get engagement, here's an eye-opening exercise. The next time you say, "The algorithm doesn't like me," replace the word "algorithm" with "audience."

Ouch, right?!? What if the truth was that you need to create better content for your followers? Content they will actually enjoy.

To gain a better understanding of what your audience responds to, go to your analytics (we discussed those when we talked about 'The Best Time to Post'). Pay extra attention to what type of content gets the most reach, comments, saves, and shares. Then, create more of it!

Sometimes, we get stuck in old ways, using outdated strategies that no longer work or we're simply burned out. Taking a break to identify what brings your audience the most joy and focusing on that is often a quick fix.

Once you have an idea of what's working, make sure you:

- Craft captions that genuinely connect with your followers (including calls-to-action whenever possible).
- Be specific with your hashtags to reach the right people.
- Prioritize quality over quantity in your posts, giving people a reason to engage with them.

At the end of the day, some posts perform well, while others won't get as many likes or views. Recognizing this takes the pressure off.

Your self-worth isn't determined by the number of likes or followers you have!

"Can I post about controversial subjects on social media?"

You really don't have to join every argument you see in your feed. When you're passionate about an issue, it can be super tempting to feel like it's your job to speak out. If you choose to, always engage in a way that presents your best self to the world.

My mom was a first-grade teacher for over thirty years, and every morning after the pledge of allegiance, she would have her first graders repeat after her: "What you put out, comes back to you."

If you decide to join a hot topic conversation because it's important to you, always speak out for what you're *for* instead of what you're *against.*

This allows you to use your voice but show your best self on social media—the one that people want to get to know and collaborate with on their next project.

Remember: What you put out comes back to you (especially on social media).

"Should I pay to 'boost' my posts?"

I know it's hard to resist when Facebook tempts you with a $5 credit to boost a post to 'get more followers.'

IMPORTANT PSA: Stop paying to 'Boost' posts of your headshot or your acting reel on social media. This is not a targeted marketing strategy and is not the best way to spend your money.

Sure, you may get a few more likes and comments on that post, and that may feel good. But what did it actually achieve?

And please, never 'boost' video content you don't own the rights to (e.g., acting clips, your reel, etc.). That's a copyright violation and can get your ads account shut down (or worse, your social media account disabled).

Save your money. Spend it elsewhere.

#MindsetMoment Connection Not Perfection

I can offer you plenty of tips for improving your stories, posts, or working with the algorithms, but one of the best pieces of advice I can give for both your social media and your acting is to let go of the pursuit of perfection.

Your videos will never be flawless, and your posts might occasionally contain typos (thanks, auto-correct).

Just post them with the intention of connecting with your followers, even if it's just one person. Experiment to see what works and what doesn't. The more you practice, the better your posts will become.

Remember, social media is social. It's about connecting with your audience.

REPEAT AFTER ME:

I will focus on connection, not perfection.

CHAPTER 6

CREATE BETTER VIDEOS ON SOCIAL MEDIA

Video: Your Secret Weapon

Video content dominates social media. Your fans and followers crave authentic interactions, and video is a powerful tool for connecting with them.

I know many of you are reluctant to create videos for social media. I get it. I was a nervous wreck when I created my first Periscope way back in the day (#RIPperiscope).

But fast forward eight years and the majority of my content is video. In this chapter we'll talk about how you can create better videos online. But first, you need to feel comfortable in front of the camera. Let's kick things off with four tips that helped me conquer my fear of video.

Think Small

Film your videos in small sections that you plan ahead of time. This is especially helpful for longer YouTube videos. If you try to shoot something all in one take, the fear of making a mistake can take away from your performance. Your video doesn't have to be perfect to be good. The goal is *connection not perfection*. A little word fumble isn't the end of the world. Your audience knows you're human. If you make a mistake, just pause, center yourself, restart from the top of that sentence, and try it again. The great advantage of recording on camera is that we can cut out the big mistakes and keep the best pieces to deliver the most impactful performance.

Write A Script Or Outline

Too often, when you try to wing it on camera (even if you really know the subject), you'll wander off-topic, ramble, or forget a crucial point you want to convey. This is especially important for videos on social media. Attention spans are short, and having an outline or script will keep you on track and your audience engaged.

If you're new to being in front of the camera in this way, I recommend starting with a full script, and as you become more comfortable on camera, move to bullet points or an outline— whatever works best for you.

Talk To Someone

Speak directly to the lens, which is a different experience from acting on camera. It may help to imagine you're speaking with someone you know, like your best friend. Address the camera as if it were a person: you're not talking to millions of people—talk to one person.

Watch Yourself

One of the best things you can do is record yourself on camera, then play back the footage to see how you appear on screen. For the first four years, I personally edited all of my YouTube videos. Although it was time-consuming, it was the absolute best way for me to become comfortable on camera. Why? Because I could watch myself, identify areas for improvement, take notes, and implement those changes in my next video. Just like anything else, being good on camera takes practice. The more you do it, the better you'll get.

Now that you're feeling a bit more at ease with video, let's talk about how you can make your videos better.

Creating Better Video Content

After analyzing my clients' best-performing content, I've identified five consistent elements that make these videos stand out:

- **Good lighting that allows us to see you clearly.** This can be achieved with a simple 3 point lighting setup, with a ring light or by utilizing natural window light.
- **A stable camera.** All you need is a tripod. Curious about my favorite tripods? They're listed in the companion guide to this book.

- **A compelling hook that grabs our attention and stops the endless scroll.** Consider this: the average person scrolls through over 300 feet of content daily on social media. That's equivalent to the length of a football field or six Hollywood signs placed end to end. It's crucial to create a strong hook, whether it's verbal, on-screen text, or captivating action or movement.
- **High watch time** (More about this in the next tip).
- **A clear call-to-action (CTA) within the post.** Think about what you want your audience to do after watching your video, or even during it. What specific action should they take? Do you want them to subscribe, follow, comment, or answer a question? You can find some CTA examples in Chapter 5 for reference.

Watch Time

The most crucial metric for your videos is watch time.

How long do people stick around before clicking away? If we lose interest within a few seconds, we're gone, and your video won't get the views you want.

Every social media algorithm highly values the '*Watched Percentage*' of videos.

That means, if you watch 5 seconds of a 5-second video, that's 100% watch time, and the platform thinks, "This video is fantastic!" and promotes it further. On the other hand, if you only watch 20 seconds of a 60-second video, that's 33% watch time, indicating to the platform that the video isn't as engaging, and it won't receive as much promotion.

That's why shorter videos go viral more frequently than longer ones. The average viral video is about 7 seconds long. Keeping viewers engaged in longer videos, especially without a strategy, can be challenging. So, grab our attention in the first few seconds and deliver your content as succinctly as possible.

Overcoming FOP

One of the common challenges new creators face is the fear of posting (FOP). You may have outstanding content, but if you're too afraid to hit the "post" button, your brilliance will never be discovered. Here are 3 tips to overcome your FOP:

If you're not posting or you're afraid to create a video because you think: *"What will people think?"*

My garden is filled with sunflowers. But, no matter how tall, vibrant, or stunningly beautiful they may be, some people will still not like sunflowers. In the same way, not everyone will resonate with your content. I get unfollowed every single day and that's completely okay with me. Your goal on social media isn't to attract all people. It's to attract the *right* people. They will love you. The wrong people won't.

Or maybe your FOP shows up with thoughts like: *"What if I suck?"*

Well, don't worry. You will. I did. Everyone did. Be brave enough to be bad at something new.

If you don't take the first step, you'll never get good at anything.

Remember, you don't need to get it 'perfect,' you just need to get it going.

Or maybe the FOP thought holding you back is: *"What if nobody 'likes' it?"*

I want you to listen closely (especially if your audience is small).

Everyone starts at 0:

0 Followers
0 Posts
0 Likes
0 Comments
You still have to create even when no one's watching or listening. It's really no different from other parts of your acting career, right?

So just show up. Because when you show up consistently, others will, too.

CHAPTER 7

YOUR NETWORK IS YOUR NET WORTH

The number one reason to use social media for your acting career isn't followers. It's networking.

Social media is the best (not to mention free) tool for creating and building relationships for your career. Before social media, relationship-building was limited to who you could meet in person. But now you can meet and nurture relationships with people you want to know online *if* you're strategic about who you want to meet and how you'll connect with them.

Beyond meeting new people, social media is also the best way to stay 'top of mind' with people you already know and the fastest way to book new work. Why? Because you can put yourself in circles of people who already know, like, and trust you without having to wait for the phone to ring.

In this chapter we'll talk about networking with people you already know. We'll get to expanding your network and making *new* connections in Chapter 8.

The Virtual Coffee Shop

Social media makes it possible for you to have virtual 'coffee dates' on your couch, in your yoga pants, every week with people who know, like, and trust you.

One of my students booked a national commercial thanks to her virtual coffee dates. Linda had been interacting on social media with a casting director who had called her in several times. An hour before she received the commercial audition request Linda had commented on one of the CD's posts. The comment wasn't about an audition, it wasn't even about acting. It was about the casting director's dog. Coincidence?

No. By consistently engaging with her, Linda was on the casting director's mind. In fact, the CD told her, "I saw your name on my post and thought, 'Linda would be perfect for this!'"

By creating your own virtual coffee shop, *you* can stay top of mind with the people who know, like, and trust your work and you can attract more auditions and opportunities for your career.

#HomeworkTime

Who do you know? Take a few minutes to write down all the people you've worked with: directors, producers, writers, fellow actors, audiobook producers, etc. You can comb through your emails, old call sheets or cast lists, too. If you're just starting out, acting coaches and actors in your classes are great contacts. You'll add more people as your career grows.

Constant Contacts

Connect your email or other social media accounts (if possible) to each platform. This ensures the social networks will give you a steady stream of 'Connection Suggestions' for anyone you meet on your *next* booking. You may also discover some contacts that you missed from the #HomeworkTime you just did.

The 3 R's of Networking: Reconnect, Remind, Rekindle

Now that you've identified the people in the business that already know, like, and trust you, look them up on your social network of choice, follow them and reconnect.

If you haven't seen someone in a while, I know it can be hard to know what to write. But social media helps you by giving you clues for conversation starters. You just need to be a digital detective.

Read their bio and look at some of their posts. Maybe they just booked something, maybe their son just graduated from college, maybe they just had a baby or maybe they just grew a beautiful tomato in their garden (it doesn't have to be life-changing).

Let's say their son just graduated from college:

'Like' and then 'Comment' on that post, saying something like, "Wow! It's been forever since _____ (give them a reminder). Congrats to your son! He was 10 the last time I saw him. So glad I found you here on Instagram again."

We **RECONNECTED**, we **REMINDED** them when we worked together, and we **REKINDLED** the relationship so we can stay in touch in the future.

Organizing Your Contacts

As your following grows, it can become increasingly challenging to see the people you really want to network with in your feed. Social media algorithms may limit your ability to see all of your contacts, hindering your ability to engage with them.

The solution is to get organized! Add the people you already know to your 'favorites' or categorize them into well-structured 'lists' on your social media profiles. By organizing your industry contacts on each platform, you'll be able to stay top of mind and consistently nurture your relationships.

Your Connections = Your Career

How well you stay in touch with your industry connections affects the number of opportunities you'll have and the trajectory of your career. Be consistent with your virtual coffee dates and keep these connections strong. As an actor, your network is your net worth.

Chapter 8

Making New Connections Using Social Media

If you know the *right* way to connect, social media is also a fantastic tool for expanding your network with the people or projects you want to be a part of. In this chapter we'll fix the biggest mistakes actors make when making new connections on social media.

Specificity Is Everything

Like spaghetti posting (in Chapter 2), too many actors have a spaghetti approach to their networking. They try to meet everyone in the business, but end up not connecting with anyone.

The first step to making new connections online is getting specific about *who* you want to network with:

- What type of work do you want to do?
- What type of stories do you want to tell?
- Who do you want to meet in this business?

You don't need to meet thousands of new people to change your career; you just need to meet and build relationships with a few of the *right* people.

Don't Forget The Little Fish

If you focus on the 'Little Fish,' social media is an amazing place to expand your network and meet new people. Let me explain…

Most actors focus their online networking efforts on the 'Biggest Fish' in the pond—the Steven Spielbergs, the biggest casting directors, Academy Award-winning directors, and their favorite celebrities. But there are lots of more accessible 'Little Fish' you can build relationships with online.

This is particularly important on social media because the 'Big Fish' may:

- Not even manage their own social media accounts.
- Have much larger audiences, making it more challenging to get their attention.
- Have verified accounts and filter their notifications and might not even see your message.

Also, don't forget your fellow actor and filmmaker friends. They're your own group of 'Little Fish.' Together, you can become 'Big Fish.'

Expand Your Audience

Casting Directors are vital to the acting industry. I count many among my friends. But your audience isn't just casting directors and they shouldn't be the sole focus of your networking efforts.

Depending on how you fit into the business there are so many people who can be important for your career:

Agents
Managers
Yes… Casting Directors
Casting Associates
Producers
Writers
Fellow Actors
Directors
Crew Members
Acting Coaches
Publicists
Lyricists/Composers
Choreographers
Audiobook producers
(And I could keep going depending on how you fit into the acting industry)

Include some people outside the casting director box and you'll have greater success with relationship-building.

Push Versus *Pull* Marketing

To effectively network on social media you need to understand the difference between 'Push Marketing' and 'Pull Marketing.'

The difference between push and pull marketing is in how you approach your customers. An actor's 'Customers' are the people who can hire them or connect them with someone who can (casting directors, agents, directors, producers, fellow actors, etc.).

Push marketing promotes products (or yourself) by pushing them onto people. Too many actors send direct messages asking people to watch their reel. They'll also ask for auditions or tag casting directors (who don't know them) in their posts. These are push marketing tactics. They're a super direct, pushy approach that rarely works on social media until you've established a relationship.

Pull marketing is the opposite. Pull marketing gets 'customers' (casting directors, agents, directors, producers, fellow actors, etc.) to seek out your brand. It pulls them in, so they come to you. How do you pull someone in and get on their radar?

Getting On Someone's Radar Online

The quickest way to *pull* someone in is through their ego. And I don't mean ego in a negative sense.

We naturally gravitate toward people who make us feel good about ourselves, right? That means when you offer praise, compliment, share, retweet, or simply follow someone online, you appeal to their ego in some way.

When you actively engage online with the people you want to work with in a real, human way you *pull* them in. And, over time, if you're consistent and authentic, they'll get curious and click over to your profile to find out more about you. It's pull marketing and it's also human nature. Think about it: Have you ever clicked over to find out more about someone who leaves thoughtful comments on your posts? Of course you have.

Here's why pull marketing is so magical. When your initial interaction with someone new is about *you,* it pushes people away; if you make it about *them* it pulls them in. Pull marketing is the most effective way to get noticed online.

Be Interested

In his book, "How to Win Friends and Influence People," Dale Carnegie said "You can make more friends in two months by becoming interested in other people than in two years by trying to get other people interested in you."

Translation: You can make more friends in two months through *pull* marketing than in two years using *push* marketing.

So, instead of starting your social networking efforts with an 'ask,' focus on building the relationship. Your networking should be based on your desire to connect with people, not on getting something in return for your efforts. Focus on social *giving* before you even think about social *getting*.

Intentional Engaging

When I tell actors to join conversations online with the people they want to meet, I'm often met with blank stares and questions like, "What do I say? What do I comment?"

SPOILER ALERT: Casting directors, agents, filmmakers, etc., are human, just like you. So, create a human connection.

Just dropping an emoji on their post isn't going to cut it. When you see their post in your feed you can:

- Authentically join their conversations.
- Congratulate them on their success.
- Add commentary that's respectful and genuine, adding value to their conversation.

- Think about what part of their post resonates with you and why. If it doesn't resonate, don't comment. There will be other opportunities.

"When Do I Tag Someone In My Post?"

In my social media consulting I help a lot of casting directors, agents, and directors with their social media. The number one thing they ask me to teach actors is:

Tagging people on social media posts they have nothing to do with is spammy and is not good networking.

Tag someone in a post when:

- They're actually in the post.
- You're giving someone a shout-out.
- The tagged person is associated with the project you're posting about.
- You're collaborating with an account.
- You're quoting someone.
- You're answering someone's question publicly.
- You're posting someone's review or testimonial.
- You're tagging a product in your post.
- You're resharing a post and want to give someone credit.

Remember: Tagging is supposed to be beneficial for both you *and* the person you're tagging.

The Temperature of The Relationship

If you've been building a relationship and authentically engaging for a while you may be wondering "When *is* the right time to ask for something?"

There's no one-size-fits-all answer to that question. Building real relationships takes time. Be patient! It may take 300 comments to make a connection with one person and only a few with another. But regardless, it all comes down to the *temperature* of your relationship.

When you meet someone, the relationship is cold, but over time, connections get warmer. Sometimes they become hot and they know your name and will actually reach out to you. Most actors ask for something when the relationship is still cold. It's rarely successful.

Even with connections you already have, you have to stay in touch to keep those relationships warm. If they haven't heard from you in five years, the chances they'll go out of their way to help you are pretty low.

Strong, lasting relationships that lead to more auditions and opportunities take time to develop. Be patient. Great relationships aren't built in a day; they are built daily!

#HomeworkTime

Nothing will happen if you read this chapter and do nothing. Read it again. Then start making connections and building authentic relationships online. Networking only works if you #DoTheWork.

CHAPTER 9

RUN YOUR SOCIAL MEDIA LIKE A BUSINESS

Now that you're using social media for your acting career, it's time to treat it like a business. That means using it with purpose, not for playtime. In this chapter, I'll share some quick tips to help you make the most of your time, so you can dive in, accomplish what you need, and get back to your true purpose—acting!

Social Media Can Be A Waste Of Time

Let's be honest. Without a plan, social media can be time-consuming and a waste of time for your acting career.

There are simple systems you can set up to make sure you stay in control of your social media and your time. For instance, choose a specific time every day to check notifications and messages and engage with fans and followers. Do it all day and you risk getting sucked into the social vortex for hours (and hating social media).

To help you stay consistent, pair the time you choose with a daily ritual like your morning coffee, lunchtime, or your daughter's TV time.

You Are The Boss Of Your Social Media

You don't need notifications constantly popping up on your phone every time someone follows you or tags you in a post. Remember, you are the boss of your social media, not the other way around! Notifications can make your phone feel like it's in control, but you have the power to decide when you want to see them. Take charge by disabling push notifications, so you only get alerts when you intentionally sign in.

Set Social Time Limits

Establishing designated times to unplug from your phone is crucial in today's hyper-connected world. It's all too easy to find ourselves constantly glued to screens, which can lead to burnout and decreased productivity.

An effective strategy is to set specific times during the day when you intentionally disconnect from your devices. For example, during meals or family time, consider placing your phone in another room, out of sight, out of reach.

By creating these unplugged moments, you can recharge, improve your focus, and maintain a healthier balance between your digital life and the real world. Remember, you control your devices; don't let them control you.

Take The INSTA Out Of Your Instagram (And All Of Your Social Media)

Just because something happened today doesn't mean you need to post about it today. Instagram's name might suggest real-time posting, but there's absolutely nothing wrong with sharing a photo from two weeks ago today. Getting into the habit of not posting in real-time can help you avoid spoilers, create more impactful posts, and maintain your privacy. No need for #TBT or 'latergram' because nobody knows when the photo was taken. It's always better to be intentional than instant.

There's An App For That

There's an app for nearly every task on your 'Social Media To-Do List,' and trust me, they make everything easier, faster, and often, way cooler. You can harness the power of apps to:

- Schedule your content in advance.
- Create captivating video and audio content.
- Discover relevant hashtags for your target audience.
- Edit your acting clips for social media.
- And more...

If you're not using apps, you're wasting valuable time! #Truth.

For my current timesaving app recommendations, check out the companion guide accompanying this book at www.socialmediaforactors.com/guide.

Don't Get Hacked!

Several times a week, I hear from an actor who's locked out of their social media account or had it deleted due to hacking. Fortunately, most hacking is avoidable if you know what you'll learn in the next few pages.

Before you skip the rest of this 'unsexy' chapter, you need to know:

- Hackers don't just target large accounts.
- Hackers usually don't know their victims.
- Having a private account does NOT protect you (re-read Chapter 1 to find out why most actors shouldn't have a private account).
- A strong password isn't full protection.
- Social media platforms don't have the time or resources to help.

Do I have your attention now?

If you are reading this book and spending any amount of time on your social media strategy, you need to protect your accounts. The number one thing you must set up today is called two-factor authentication. If you already have it set up, high five! You rock! Skip to the next chapter.

If not, keep reading…

A Door With Two Locks

Two-factor authentication means there are two factors that must happen before someone can access your account.

Technically, when you set up a social media account, you have single-factor authentication, single-factor meaning that to log in, you have to enter your password. One single step to verify that you are who you say you are. The problem is that passwords are easy to hack, even if they're strong. So if this is your only method of keeping your account safe, it's like you're leaving the keys in the door.

Two-factor authentication adds an additional layer of security by making it harder for hackers to gain access to your account. Even if your password is hacked, a password alone isn't enough to get into the account. The social network will also send a special access code to your phone that you must use to get in. Most hackers don't know their victims, so they don't have your phone. It's double security.

It's like a door with two locks. Your password unlocks the lock and the access code unlocks the deadbolt. They need BOTH to get into your account. Two-Factor authentication = Two factors to get into your account.

If growing, connecting, and thriving online is a goal for your social media, you must do this today on all of your social media accounts.

#HomeworkTime

Set up two-factor authentication right away! Need help? The companion guide for this book provides step-by-step instructions for you.

Chapter 10

Handling Challenges Online

Navigating social media comes with its fair share of challenges. Dealing with bullies, trolls and burnout can be part of the journey. This chapter gives you valuable insights and strategies to help you overcome some of the most significant challenges you'll face online.

A Highlight Reel

Be mindful of the content you consume on social media. People tend to share all the amazing moments in their lives and often leave out the lows. This can lead to a 'compare and despair' mentality. I always remind myself that what I see on social media is mostly a highlight reel.

Take a moment to go through your social media feed and identify what makes you feel good and what doesn't. Then, consider unfollowing, unfriending, or muting accounts that upset you, consume too much of your time, or make you feel less positive about yourself. It's okay to unfollow accounts that no longer contribute to your algorithm of happiness.

Stop Comparing Yourself To Strangers Online

I know this one is hard, especially if you also compare yourself to other actors at auditions. Please stop comparing yourself to others, both offline and online. Remember: You can't compare your Chapter 1 with someone else's Chapter 10. The accounts you're comparing yourself with may have been on Instagram since 2011, or you're comparing your 'likes' to someone with 100,000 followers. What's important is that you're working to improve your social media for your career, not to have more followers or likes than the accounts you see in your feed.

If you struggle with this, consider turning off your 'likes' on platforms like Instagram. You won't see your 'likes' or other people's 'likes' when scrolling through the feed. This small change can stop any 'compare and despair' that's based on vanity metrics.

#MindsetMoment

REPEAT AFTER ME:

I create authentic content free of the pressure to be like anyone else!

Set Up Filters For Words You Don't Want To See Online

On most platforms, you can mute certain words you don't want to see on your pages. When you do this, they won't appear in your notifications, feed, comments, or direct messages. You can set this up right in your privacy settings (it's called something different on every platform: look for 'Hidden Words,' 'Mute Words,' or 'Filter Keywords'). If you're dealing with online bullies, this simple step can have a huge effect on your mental peace.

Don't Be The Bully

Always be kind online. If you don't agree with someone, consider just unfollowing them. Don't leave hateful comments or try to teach them a lesson. Remember my golden rule: Tweet others the way you want to be tweeted! (It's still my golden rule, even though it's not called Twitter anymore).

Face Your Problems, Don't Facebook Them

I know this business can be tough, and staying positive can be a challenge. But remember, social media is not your diary. The fastest way to get unfollowed is by sharing negative energy on people's news feeds. If you need to rant or vent, consider using a journal or confiding in a friend. Social media isn't the place for that.

Personalizing your social media doesn't mean sharing totally personal things. Remember: Your account is part of your first impression for your acting career.

Avoiding Social Media Burnout

We've all had a favorite account or a friend suddenly go silent, disappearing from our feed for weeks, maybe even months.

Burnout is a real issue for many online creators. In this section, I'll share three tips that have helped me avoid social media burnout.

1. Social Media Success Is A Marathon, Not A Sprint

Everyone craves quick success, and we often hear about 'overnight success' stories of an actor's single viral video propelling them to influencer status. First of all, it's entirely normal to wish for an easy path to success, but it's not the typical online reality. These 'overnight success' stories are rare. Achieving success on social media takes hard work and patience. Keep pushing forward, my friend. Growing a real, engaged audience on social media is a marathon, not a sprint.

2. Embrace Failing Forward

Growing an audience on social media takes patience and the continuous practice of what author John C. Maxwell calls 'failing forward.' Failure is essential for future success. Every successful online creator has encountered numerous learning opportunities (aka failures) on their journey to where they are today.

When you learned to walk, you probably fell *a lot* before you figured it out. Similarly, there are many valuable lessons to be learned with each social media post. These lessons contribute to you becoming a creator who consistently produces engaging content.

Examine your social media accounts and identify areas that could benefit from improvement. Is it your photos, captions, or perhaps your videos? Whether it's video structure, hooks, or editing, choose one aspect and start making changes to enhance it. The more you practice, the better you'll become.

3. Always Have 'Content On Demand'

 If you rely on coming up with ideas on the fly, you'll find yourself stuck on the endless content hamster wheel and risk eventual burnout. I make it a habit to always have content on demand ready to go, ensuring that I never have to say, "I don't have anything to post today."

There are numerous posts an actor can have, queued up on their phone or in a social media scheduler. Look on your phone and computer for:

- Throwback photos of you performing.
- Photos that document your journey as an actor (your first acting class, first show, first time in the booth, and any photos or videos celebrating career milestones).
- Headshots (current, your first, funny ones).
- Photos/Videos from acting projects that haven't been released or aired yet.

All of these are types of content you can have at the ready for those times when you're busy or simply not in the mood to create new content. Having content on demand is my long-term secret to avoiding burnout.

#MindsetMoment

REPEAT AFTER ME:

I celebrate all the tiny victories I make online everyday!

It can be easy to only focus on the major milestones, but don't forget to celebrate the small wins you make every time you show up for yourself online.

Chapter 11

Social Media Tips For Introverted Actors

Hi there, introverts! I see you. And guess what? I'm just like you. That's why I've dedicated an entire chapter of social media tips to your unique needs.

Now, don't let the word 'social' intimidate you. I firmly believe that social media is an introvert's paradise because it offers a platform for short, purposeful social interactions on your own terms. Once you know how, introverts can leverage their strengths and truly flourish on social media. I'm living proof of that!

In this chapter, I'll share five strategies that have not only helped me conquer my #OnlineShyness but also propelled me to achieve significant success on social media.

The First Step To Unlocking Success Lies In Tweaking Your Mindset

Instead of feeling anxious about "talking to people online," I shift my perspective to "I'm going to brighten someone's day today!"

We all appreciate receiving heartfelt comments on social media, don't we? So, why not make someone else's day?

With this mindset, it is easy for me to identify a hashtag or conversation that I'd like to join and connect with new people.

In fact, I came to the realization that it was somewhat selfish of me not to engage with others the way I appreciate them engaging with me. Whenever someone took the time to comment or message me, expressing their love for my post, it genuinely made my day. So, why hadn't I been paying it forward? Let's spread the love and make someone's day!

Establish A Consistent Routine

Routine can be a game-changer. It not only simplifies things but also gives you boundaries. For instance, if I make it a habit to engage between 7:30 and 8:00 pm, I can commit to that time, knowing that I can put my phone down and recharge afterward. This is especially important for introverts. Open-ended engagement time can be overwhelming. Remember the consistent schedule we discussed in Chapter 9 for managing your social media? Stick to it, and you'll appreciate the difference it makes.

Don't Strive For Perfection

Introverts often spend a significant amount of time in their own heads, and that can make their inner critics particularly vocal. If you catch yourself constantly asking, "Is this good enough?" or doubting whether your followers will care, you're likely overthinking it. Don't let fear hold you back. Aim for connection, not perfection, and just go ahead and post. If you're struggling with the fear of posting (FOP), revisit the end of Chapter 6 for tips on overcoming it.

Listen Before You Speak

Social media feeds can be quite noisy, with everyone trying to make their voices heard. But if you take a moment to pause and listen, you'll gain insights into what people truly want, allowing you to tailor your posts to help them. Introverts often excel at the art of listening before speaking. This can be your superpower on social media.

Find Your People Online

Many people think introverts are always shy. But the truth is, introverts can shine brilliantly when discussing topics they're passionate about. Seek out *your* community and conversations that interest you.

You'll find groups on nearly every platform, whether they're Facebook or LinkedIn groups, or communities of individuals using specific hashtags. Engaging in discussions that align with your interests makes it easy to connect with new people.

Chapter 12

Growing Your Audience

If you've jumped ahead to this chapter, I see you!

But here's the deal: If you don't put into practice the concepts from the previous chapters, you'll struggle to grow on social media.

I understand your eagerness to increase your following, but let's get something straight—that's not your first step.

Here are the steps I teach my clients (in order) for successfully building an audience online:

STEP 1: Figure out how to tell your unique story online.

STEP 2: Craft a compelling first impression that reflects your story across your usernames, bios, highlights, and more.

STEP 3: Develop a posting plan that aligns with your story.

STEP 4: Identify hashtags that not only resonate with your story but also place you in conversations with potential followers who will genuinely connect with you. And *only* after you've tackled steps 1-4:

STEP 5: Implement engagement strategies to foster growth.

STEP 6: Collaborate with like-minded accounts to further expand your reach.

I understand your desire to skip ahead but, do yourself a favor. Do the work, then meet me back here when you're ready.

Why A Following Really Matters

Let's address the elephant in the room, shall we? It's true that having a significant number of followers can sometimes improve your chances of getting cast.

At a recent film festival panel, this discussion came up and one of my fellow panelists, an indie film producer, mentioned that for foreign distribution he always had to provide follower counts for any actor "above the line" (the major creative talent, including the director, actors, writers, and producers). Having a cast with more social clout, which essentially translates to free advertising, could result in negotiating thousands of dollars more in foreign distribution deals. So yes, having a very high number of followers can, in some cases, enhance your castability.

However, I'd like you to look beyond that, because there are plenty of reasons why having a following matters beyond getting cast.

Having a built-in audience can help you:

- Promote your friends' achievements.
- Advocate for causes and charities that matter deeply to you.
- Promote the content you're creating.
- Garner attention for your crowdfunding campaigns.
- Boost your visibility and influence.

Plus, if you also teach acting or voice, or offer headshot or reel services, having a social media audience can provide you with a built-in customer base for your business.

In short, it's not just about followers; it's about growing an audience for your projects and the things you're passionate about in your life.

Next, let's talk about the common mistakes actors make when attempting to grow their audience. Then, we'll delve into practical strategies to help you grow yours.

Your Followers Aren't Just Numbers On A Screen

You're not growing numbers on a screen, you're cultivating a community! Behind those digits, your followers are people just like you, each with unique struggles, aspirations, and dreams.

Prioritize engagement with the fans and followers you already have. Show them your appreciation. Create personal connections.

Consider this: If you have 400-500 followers, you've essentially filled a Boeing 747 airplane. With 3,400 followers, you've got an audience comparable to the size of the Oscars crowd. And if you're at 17,500 followers, that's like filling the Hollywood Bowl. That's a lot of real people.

#HomeworkTime

Dedicate 5 minutes today to engage with the comments on your posts and start conversations with some of your followers. Visit a few of their accounts and engage with their content, making meaningful connections!

Buying Fake Followers

When you focus on your follower count as a number you make poor decisions about its growth. Some actors hear that having a larger number will lead to bigger roles, so they resort to boosting their numbers by buying fake followers.

But, fake followers don't contribute value to your community or career. They're essentially empty accounts, despite what some websites may claim. Fake followers don't engage with your posts.

They're just numbers on the screen, devoid of any real human interaction. As a result, you might end up with hundreds of thousands of followers but little to no engagement, which negatively impacts your visibility on any platform.

If you're cast based on your social media numbers, it's because you bring a pre-existing audience to the project, whether it's 50,000 people or 50 million. But, keep in mind that it's not your follower count that helps you get cast; it's the engaged audience you've cultivated.

In the end, people want to collaborate with people they know, like, and trust. Buying followers undermines that trust and potentially damages valuable industry relationships.

#MindsetMoment

REPEAT AFTER ME:

My follower count does not define my value as a person.

I am enough. I am valuable beyond what any external vanity metric on a screen says.

Success Leaves Clues

To grow your social media you must make it a habit to:

- Learn from your success.
- Surround yourself with success.
- Collaborate with success.

While there's no quick-fix solution for instant growth, these three steps are the winning formula for growing your audience faster. In the next few pages, we'll discuss them all.

Learn From Your Success

As a former social media manager, I love analytics because they tell you exactly what your audience is enjoying, helping you create better content. Knowing what they like is the difference between a stalled account and one that goes viral. Creating more shareable content that resonates with your followers helps you attract the right potential followers.

Open up your analytics: Not sure where to find them on your social network of choice? The companion guide to this book will help. You can download it at www.socialmediaforactors.com/guide.

Make sure you change the search to look at posts beyond last week. Depending on how often you post, you'll need to expand the search results to 6 months to a year. Also, explore different filter options to see which content received the most shares, comments, or views. Take out a piece of paper and write down:

What worked? What did people respond to? Is there a recurring theme? Maybe every time you talk to the camera and take us behind the scenes, your audience loves it. Perhaps your new series with voice-over tips is resonating with them. Maybe your 'Theatre Kid POV' video was a hit. Identify successes and patterns.

When you find something that's working and resonating with your audience, DOUBLE DOWN on that topic or type of content. If you've had content gain traction, don't create something new— create more content on that same topic!

If you create educational content, and one topic performs well, consider using Google, YouTube search, or answerthepublic.com to find other questions people have on that topic. DOUBLE DOWN on that subject. Maybe people loved the 'Home Studio Tour' of your vocal booth. Perhaps you could create another video in your home studio focusing on your microphone. What else can you teach us or share about that topic?

While you're searching for clues, also ask yourself if the post can be repurposed in a different way. If it was a 'day-in-the-life' style post, maybe you create another one by telling a story about one aspect of that day instead. Can you present the same topic differently?

If it's a video, look at the comments on your more popular videos (filter by 'views' to find them). You may find questions or comments that can inspire new videos. This is proven content— questions your audience wants answers to. Creating videos based on comments is a powerful way to connect with your audience and make them feel heard. It's also a great way to have your audience generate content ideas for you.

Lastly, look at your successful posts and videos and ask yourself if you can turn them into a series. That 'Theatre Kid POV' video that performed well can be transformed into a Part 1, Part 2, and so on.

Take some time today to dive into your analytics and learn from your successes.

Surround Yourself With Success

You've been posting great content, but you might be wondering, "Where are my followers? Why aren't more people hitting the follow button?" Well, sometimes you need to put yourself right in front of your potential followers for them to discover you. I like to call this strategy "Surround Yourself with Success."

Here's how to kick things off:

1. Create a list of five to ten prominent creators who share your target audience. Start by following them.

2. Make engaging with their content and leaving thoughtful comments a regular habit. Essentially, you're immersing yourself in the world of successful accounts that already cater to your topic or audience. Imagine this: If your account is all about your love for ice cream, then doing this is like hanging out outside an ice cream parlor, knowing that the folks going in there already love ice cream. It's about joining the community where your potential followers are. When you genuinely participate in their conversations, they'll start noticing you and click over to your account. Surrounding yourself with successful accounts also helps train the algorithms about who your account is for and what your account is about.

3. You can also use these newly discovered successful accounts for content research–learn from *their* success! Examine their content. Pay close attention to the posts that performed exceptionally well– this is proven content that your audience enjoys. Save posts that inspire you and can spark ideas for your own content. Think about what you can bring to

the table that's unique or better. What fresh perspective or viewpoint can you offer?

Now, I'm not advocating copying posts word-for-word and becoming a carbon copy of someone else. I'm talking about drawing inspiration from these creators. There's a wealth of knowledge to be gained from people who are already successful. This approach gives you tons of post ideas and helps you shape your vision as a content creator.

Collaborate With Success

The last (and often fastest) growth strategy is to collaborate with success!

Why are collaborations a faster way to follower growth? Think about it—a collaboration involves another account 'recommending' you to their followers, and there's nothing more powerful than a recommendation from someone you trust. If done right, with one post or video, you can get in front of a lot of like-minded followers.

There are many types of collaboration strategies:

- Account Takeovers
- Going Live with other accounts
- Shout-outs
- Collaborating on posts or videos
- Contests
- Sponsored post collaborations

If your goal is growth, the most important thing to consider when finding accounts to collaborate with is that they have a like-minded audience. You might already have a few accounts in mind to collaborate with, like fellow actors, cast members from your show, or your coaches. As you apply the principles from the previous section (Surround Yourself With Success) you'll discover lots of potential collaborators who share your passion for the topics or interests you post about. Keep a list as you conduct your research, foster those relationships and, in time, collaborate. Remember, you don't have to grow alone.

Collaboration > Competition

Stop viewing other actors on social media as competitors. Instead, follow them and become part of their communities. They can potentially become your collaborators, allowing you to create joint content and promote each other's accounts. On social media, choose collaboration over competition. Always.

CONCLUSION

Congratulations, you made it! You now have over 100 essential tips to help you achieve success on social media. Before you go to #DoTheWork, here's one more tip for you:

This industry requires us to use more and more technology every day. It's not going away. The better you are with these new tools, the more confident and successful you'll be.

Saying, "I'm not savvy with tech and social media," is keeping you stuck and hurting your career. Be careful about what you say to yourself because, as Oprah says, "You become what you believe." Every morning you wake up, you make a choice:

- I will or I won't
- I can or I can't
- I am or I'm not

So, instead of "I'm not good at social media," change your script to:

"I'm becoming a more social savvy actor." Because if you've read this book, you are.

I'd love to know what your favorite takeaway was from this book.

Please connect with me on Instagram.

I'm @Marketing4actors.

See you on social!

Heidi

25 POST IDEAS

FOR ACTORS

Looking for those 25 post ideas I promised you in Chapter 2? You found them! Use these ideas to inspire your next feed post, video, or story. Don't forget to include a call-to-action whenever possible.

Post #1:

Post about a performance that inspires you. Don't forget to tag the performer. Remember to include a call-to-action and ask your followers what performance inspires them.

Post #2:

Give us a backstage pass to your #actorslife. What did you do today or this week? Did you self-tape, read a script, go to class, audition, etc.? *Don't be a spoiler. Remember back to Chapter 3. Don't mention any information about scripts, characters, or auditions!

Post #3:

Recommend someone to your followers and tell us why we should follow them.

Post #4:

Post a current Headshot and say:

"Thinking I'd be a great addition to the next season of _____ [a show that you'd love to be on that fits the vibe/essence of your headshot]. What do you think?"

Make sure you tag the headshot photographer & makeup artist if you had one.

Post #5:

Post a picture of your younger self. Tell us 3 things you'd tell them if you could. (This could be great for #MondayMotivation, #TBT, #FBF)

Post #6:

Post a photo or video showing your skills in action. (For example, you singing, acting, dancing, doing accents, stunt work, special skills, etc.)

Post #7:

Thank someone in the industry for an article or video they wrote. Tell your followers why it's awesome! (don't forget to tag the creator)

Post #8:

Behind the project! Post a clip from your reel, a short clip from a performance you're proud of, a screengrab from a project or an audiogram from a recent voiceover or audiobook project. Tell us why this project was special, what you learned or a story about that day on set/in the booth/or someone you got to work with.

Post #9:

Teach us something! There are so many things you do as an actor that the average person (or actor who doesn't work in your niche) doesn't know about. Looping, pickups, put-in rehearsals, Sitzprobe, Tech Week, etc. Have to do pickups today? Post about the 111 pickups you had to do and tell us what the heck that even means.

Post #10:

Tell us what you're currently binge watching! Ask your followers if they've seen it or ask us what we're watching.

Post #11:

Share a memorable audition or performance story. This can be funny, inspiring, or even embarrassing. It's completely up to you!

Post #12:

Guess what may happen in our industry! Award show winners, movie casting, plot lines of favorite shows, etc. Ask your followers what they think.

Post #13:

What's your dream role? Tell us or show us in a photo. Or use video to perform something from that show or in the style of that show.

Post #14:

Tell us why you became an actor (in a quick video or a story in the caption). If your followers are mainly actors: Ask them when they knew they wanted to be an actor?

Post #15:

Where do you feel most alive or like to go when you need some inspiration? Maybe it's a lake, a museum, the beach, your favorite store, or your kids' room. Wherever it is, snap a photo or record a video in that spot and share why and how it inspires you. Short on time? Just search your phone for a photo of you in that place. REMEMBER: Take the Insta out of your Instagram and all of your social media. No one will know when it was taken.

Post #16:

Post about a cause you care deeply about. Tell us why it's important and, in the caption, how we can help.

Post #17:

Reintroduce yourself! If you're doing the work from Chapter 12, your audience will grow. It's great, from time-to-time, to re-introduce yourself to your followers! Here's an example:

"Hi! I see some new faces around here, so let me reintroduce myself. I'm _____ [your name]. I'm an actor who loves _____, _____, and _____ [list a few other values/interests you post about]. I'd love to get to know you better as well. Tell me 3 things you love in the comments…"

Post #18:

Share news or industry trends! Talk about an article you read recently that relates to acting (or another topic you post about). Maybe it was about a show you'd like to be a part of, trends in the industry, or a new way of thinking about the topic. Talk about it and share your thoughts. This can be in the caption or a quick talk to camera video.

Post #19:

Post about achieving a career goal or milestone! Talk about the struggle as well as the achievement in the caption. Ask your followers to share their most recent success in the comments.

Post #20:

Did you reach a milestone on your account? 500 followers? 1000? 20,000? Thank your followers in a post.

Post #21:

Create an #AuditionToScreen post! Show a small section of your self tape with the same section of the final scene. An alternative is "From Voiceover Audition to Final Spot" where you show us a small part of the recording and then show us the same section of the final spot. *You'll find examples of this post in the Companion Guide. Download it at www.socialmediaforactors.com/guide

Post #22:

Post a selfie announcing one thing you'll accomplish today. Ask your followers to post what they'll accomplish in the comments.

Post #23:

Create a GRWM video (get ready with me) to show a day-in-the-life on set, backstage, or in the booth. Only share it when it's okay to do so!

Post #24:

Post the inspiration that got you through the week. Was it a tiring 8-show week? A long day on set? What got you through it? Music, yoga, meditation, your kids? Let us know!

Post #25:

Who has made a difference in your life? A teacher? A fellow actor? Give them a shoutout on social media and tag them. Ask your followers to tag someone who's changed their life for the better.

Happy Posting!

Additional Resources

Since social media and the internet move at a lightning pace, I've limited links in this book. You can find up-to-date apps and links by downloading the companion guide for this book.

Companion Guide: www.socialmediaforactors.com/guide

*There are more than 100 additional tips, apps, and exercises in this companion guide.

Additional tutorials can be found on:

My YouTube Channel: www.youtube.com/@heididean

My Blog: www.marketing4actors.com

Online Social Media Courses for Actors:

https://marketing4actors.com/our-classes/

About the Author

HEIDI DEAN is widely known as the industry's top social media marketing expert for actors. A graduate of NYU's Tisch School of the Arts, she was a working actor for over 20 years and a proud member of SAG-AFTRA and AEA.

She founded Marketing4Actors.com, which helps actors navigate an increasingly digital world so they can build their audience, make connections, and make an impact on social media. Often called "The Wikipedia of Social Media," Heidi combines her years of industry experience with cutting edge social media strategy to help actors create more career opportunities.

Her clients include Emmy Award-winners, Broadway stars, series regulars, voice actors, audiobook narrators, casting directors, producers, filmmakers, content creators and businesses.

She currently lives in New York with her husband, Philip and daughter, Tallulah.

Find her on Instagram @Marketing4Actors.

www.ingramcontent.com/pod-product-compliance
Lightning Source LLC
Chambersburg PA
CBHW060332130626
46553CB00003B/986